# HUMPBACK WHALES

## A MIGRATION STORY

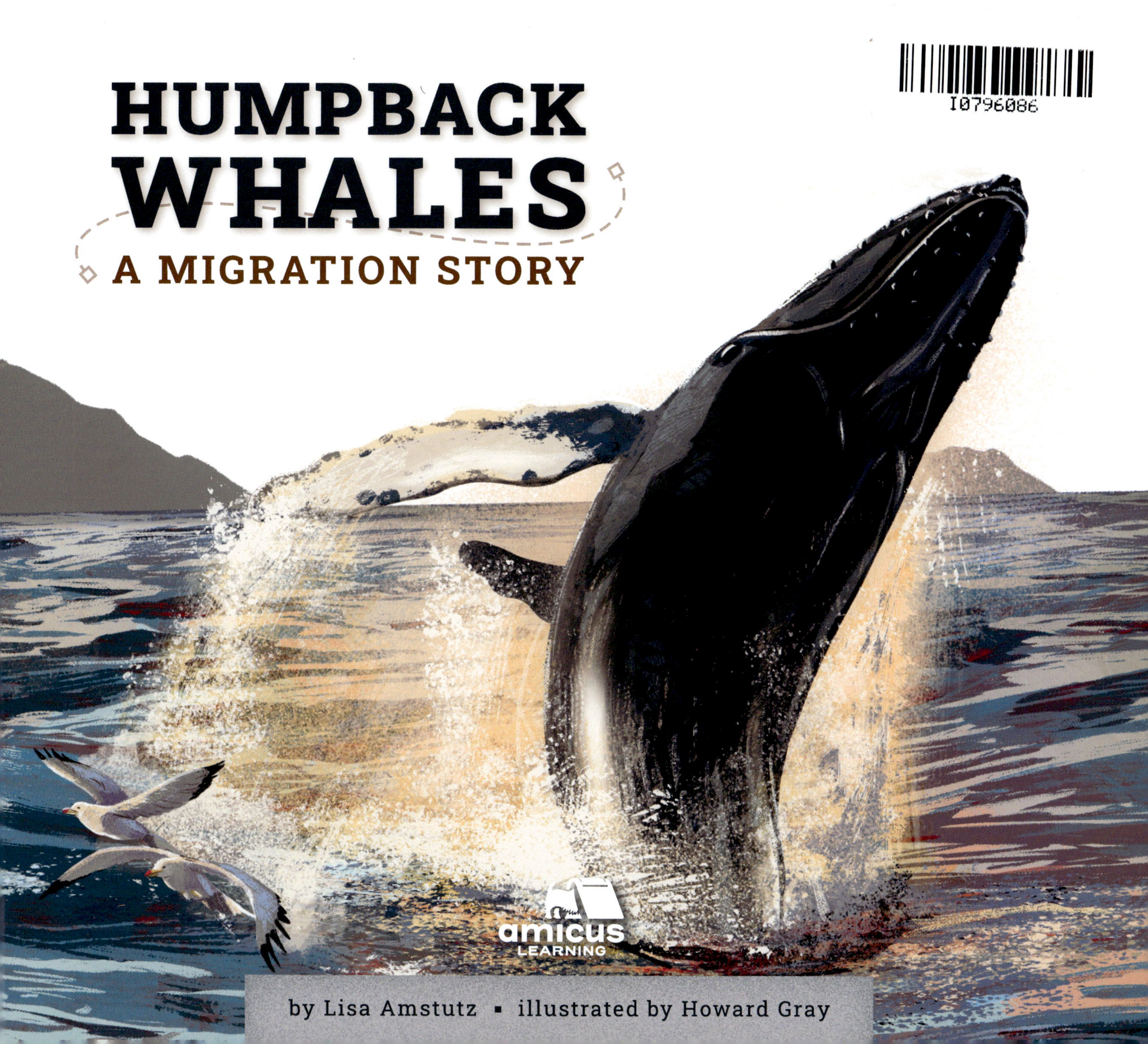


by Lisa Amstutz ▪ illustrated by Howard Gray

AMICUS ILLUSTRATED is published by
Amicus Learning, an imprint of Amcius
P.O. Box 227, Mankato, MN 56002
www.amicuspublishing.us



Editor: Alissa Thielges
Series Designer: Kim Pfeffer
Book Designer: Emily Dietz

Library of Congress Cataloging-in-Publication Data
Names: Amstutz, Lisa J., author. | Gray, Howard, illustrator.
Title: Humpback whales : a migration story / by Lisa Amstutz ; illustrated by Howard Gray.
Description: Mankato, MN : Amicus Illustrated, [2026] | Series: Incredible migrations | Includes bibliographical references. | Audience: Ages 5–10 | Audience: Grades 2–3 | Summary: "Follow the incredible journey of a humpback whale's long-distance ocean migration in this narrative nonfiction picture book that will delight animal and nature lovers and support life science education. Includes a migration map, tips to protect animals and habitats, a glossary, and further resources"—Provided by publisher.
Identifiers: LCCN 2024043397 (print) | LCCN 2024043398 (ebook) | ISBN 9781645492832 (library binding) | ISBN 9781681528076 (paperback) | ISBN 9781645493716 (ebook)
Subjects: LCSH: Humpback whale—Juvenile literature. | Humpback whale—Migration—Juvenile literature.
Classification: LCC QL737.C424 A65 2026 (print) | LCC QL737.C424 (ebook) | DDC 599.5/25—dc23/eng/20250107
LC record available at https://lccn.loc.gov/2024043397
LC ebook record available at https://lccn.loc.gov/2024043398

**About the Author**
Lisa Amstutz is the author of more than 150 children's books. A former outdoor educator, she holds degrees in biology and environmental science. Lisa enjoys learning fun facts about science and sharing them with kids. She lives on a small farm with her family.

**About the Illustrator**
Howard Gray has illustrated a selection of fiction and non-fiction children's books. He has always considered himself an artist, but with a PhD in dolphin genetics, he has a background in zoology. He is now pursuing his dream career in children's illustration from the picturesque city of Durham, UK. Find out more at www.howardgrayillustrations.com.

It is fall in Alaska. A humpback whale is on the hunt. She swallows millions of krill to prepare for her 3,000-mile (4,800-kilometer) journey ahead. She must swim to her breeding grounds near Hawaii.

Feeding is a full-time job for the whale. She blows a wall of bubbles. It traps tiny fish. She lunges through the bubbles with her mouth open wide. Her big tongue forces the water back out.

Plates of baleen keep the fish in her mouth. She eats up to 3,000 pounds (1,360 kilograms) of food a day. She stores extra energy in her blubber for the long trip.

The water grows colder as winter nears. Food is harder to find in the icy polar sea. It is time for the whale to migrate.

She gathers with one or two other whales. The pod starts swimming toward the equator. The trip will take a month or more.

The pod swims slowly. They call to each other. The whales slap the water with their tails and flippers. They do not stop to eat. On and on they swim. They travel up to 100 miles (161 km) a day.

*Whirr!* A motor drones. The noise makes it hard for the whales to hear each other. And watch out! Ships can run into whales and kill them. Luckily, this one misses them.

Other dangers lurk in the sea, too. Humpback whales get tangled in fishing nets. Orcas and sharks hunt them as well.

Some dangers are harder to see. Tiny bits of plastic float in the ocean. Krill swallow these microplastics. The whale then eats the krill. The plastic may fill her stomach. Her food may have fewer nutrients.

By December, the whale reaches Hawaii. The water is warmer here. She hears male humpbacks singing. Their songs can last a half hour. They can be heard 20 miles (32 km) away. The males breach, splash, and fight.

But the female is not looking for a mate this year. She has another job to do.

The whale heads for shallow waters. With a great push, her calf is born. It is 15 feet (4.6 meters) long. It weighs 1,500 pounds (680 kg). She pushes it to the surface to breathe.

The calf nurses from its mother. Her milk is rich in fat. The calf grows fast. It must be strong enough to travel home.

The whale takes good care of her calf. She feeds it and keeps it safe. But they must find food soon. There is not much to eat here. In April, they begin their long trip home. Mother and calf stay close. They touch flippers often. They call softly to each other.

In May, they reach polar waters. The whale teaches her calf to find food. Soon it will be able to hunt on its own. But she will stay near until next year. Then they will make the long trip south . . .

together.

# HUMPBACK WHALES MIGRATION

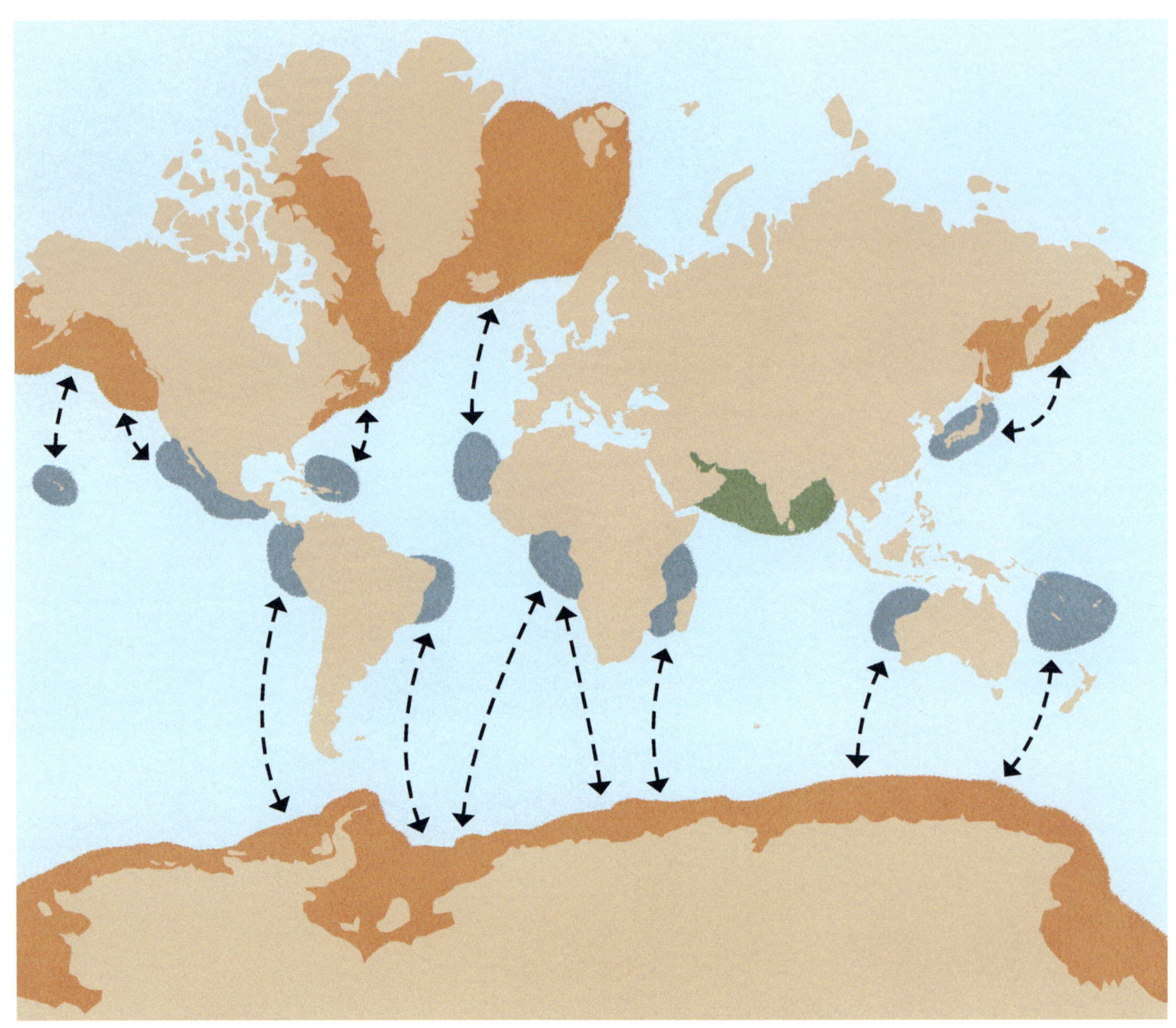

Breeding range

Non-breeding range

Migration routes

Non-migrating

## HOW CAN I HELP HUMPBACK WHALES?

Plastics in the ocean pose a big threat to whales. Here's how you can help.

- Buy items with less packaging.
- Choose products that can be recycled or composted.
- Do not litter. Trash may wash into waterways and end up in the ocean.
- Bring reusable bags to the store to carry your items home.
- Take part in a river or beach clean-up.

# Glossary

**baleen** Plates that hang from a whale's upper jaw to filter prey from the water.

**blubber** A thick layer of fat on a sea mammal.

**breach** When a whale leaps out of the water and then falls back in.

**breeding ground** An area where animals gather to find mates and raise their young.

**equator** An imaginary line around the middle of the earth.

**krill** Small shrimp-like animals that live in the ocean.

**mate** The female or male partner of a pair of animals.

**microplastic** A tiny piece of plastic.

**migrate** To move from one place to another to breed or find food.

**nutrient** A substance in food that the body needs to grow.

# Read More


Jenner, Caryn. ***Journey of a Humpback Whale.*** New York: DK Children, 2023.

McDougal, Anna. ***Humpback Whale Migrations.*** New York: Gareth Stevens Pub, 2022.

Scheffer, Janie. ***Humpback Whales.*** Minneapolis: Bellwether Media, Inc., 2025.


# Websites

**Hawaiian Islands Humpback Whale National Marine Sanctuary**
https://hawaiihumpbackwhale.noaa.gov/learn/kids-corner.html

**Humpback Whale Facts**
https://www.natgeokids.com/uk/discover/animals/sea-life/humpback-song/

**Humpback Whale Facts for Kids**
https://brisbanekids.com.au/humpback-whale-facts-for-kids/